YOUR KNOWLEDGE HAS VALUE

- We will publish your bachelor's and master's thesis, essays and papers

- Your own eBook and book - sold worldwide in all relevant shops

- Earn money with each sale

Upload your text at www.GRIN.com
and publish for free

Bibliographic information published by the German National Library:

The German National Library lists this publication in the National Bibliography; detailed bibliographic data are available on the Internet at http://dnb.dnb.de .

This book is copyright material and must not be copied, reproduced, transferred, distributed, leased, licensed or publicly performed or used in any way except as specifically permitted in writing by the publishers, as allowed under the terms and conditions under which it was purchased or as strictly permitted by applicable copyright law. Any unauthorized distribution or use of this text may be a direct infringement of the author s and publisher s rights and those responsible may be liable in law accordingly.

Imprint:

Copyright © 2016 GRIN Verlag, Open Publishing GmbH
Print and binding: Books on Demand GmbH, Norderstedt Germany
ISBN: 9783668385870

This book at GRIN:

http://www.grin.com/en/e-book/351992/security-against-chosen-plaintext-attacks

Matthias Himmelmann

Security against Chosen-Plaintext Attacks

GRIN Publishing

GRIN - Your knowledge has value

Since its foundation in 1998, GRIN has specialized in publishing academic texts by students, college teachers and other academics as e-book and printed book. The website www.grin.com is an ideal platform for presenting term papers, final papers, scientific essays, dissertations and specialist books.

Visit us on the internet:

http://www.grin.com/

http://www.facebook.com/grincom

http://www.twitter.com/grin_com

Security against Chosen-Plaintext Attacks

Matthias Himmelmann

Discourse on November 17, 2016, for the cryptography seminar at FU Berlin

1 Flashback

Up to this point, we have only discussed security definitions, where the adversary *Eve* only passively eavesdrops on a ciphertext sent between the honest parties A and B that share a key k. The following figure illustrates this issue.

Figure 1: Previous Eavesdropper Model

We have already seen some possibilities to realize this communication model, for example *One-Time-Pads* or *Pseudorandom Generators*. In this discourse, I want to introduce you to a stronger security definition, where the adversary gets the ability to learn additional details about the messages that are being send between the honest parties A and B. In the following, I am going to explain, why it is necessary to view this communication model and how to maintain security for the communication.

2 Chosen-Plaintext Attacks

Let us assume, the adversary could influence what messages the honest parties encrypt. The honest parties A and B, sharing a key k, would then proceed to send those influenced messages in encrypted form over a channel, the adversary can observe. In the below figure, I try to illustrate, what is happening, if the eavesdropper *Eve* makes A encrypt several messages $m_0, m_1, \ldots$ using k.

Figure 2: The eavesdropper influences the encrypted messages

At a later point in time, the adversary observes a ciphertext which belongs to an unknown message m. Let us even assume, the adversary knew, that the message belonged either to m_0 or m_1. Security against so-called *chosen-plaintext attacks*, or *CPA* in short, means that even in this case the attacker cannot tell, what message has been encrypted with a significantly better chance than simply guessing.

Now, a question worth asking is: Why is this a realistic security concern and how can an eavesdropper possibly influence, what is being encrypted? In the following real world examples, these questions are answered.

Example 1. Let's assume the adversary was typing on a terminal in a computer, which then encrypts and sends everything the attacker writes using a key shared with a remote server, thus unknown to the attacker. Here the adversary controls exactly, what is being encrypted, but the encryption scheme should stay secure when it is used to encrypt data for another user.

Example 2. In World War II a famous example for chosen-plaintext attacks took place. US cryptoanalysts intercepted an encrypted message from the japanese, which they were able to partially decode. It stated that the Japanese were planning an attack on AF, where AF was a ciphertext, the US was unable to decode.
The US now believed, Midway Island was the target, but couldn't convince the authorities of this assumption, since the general belief was that Midway Island could not possibly be the aim.
The US then carried out a *chosen-plaintext attack* by encrypting the fake message that Midway island was low on water supply. The Japanese intercepted this message and immediately reported to their superiors "AF is low on water."
Of course, this was the proof the US cryptoanalysts needed and the US immediately sent several aircraft carriers, resulting in the rescue of Midway Island.
If the Japanese encryption scheme had been secure against CPA, this strategy would not have worked and history might have turned out very differently.

Although we have now accepted the necessity of CPA security, we still have to properly point out, what it means for an encryption scheme to be secure against chosen-plaintext attacks. Consequently, to formally define *security against chosen-plaintext attacks*, let's consider the following experiment defined for any encryption scheme $\Pi = (Gen, Enc, Dec)$, adversary A and value n for the security parameter.

The CPA indistinguishability experiment $\text{PrivK}_{A,\Pi}^{cpa}(n)$:

1. $Gen(n) \leftarrow k$ generates a key k of length n.

2. The adversary A is given the length n as input and oracle access to $\text{Enc}_k(\cdot)$. He outputs messages m_0, m_1 with $|m_0| = |m_1|$.

3. $b \in \{0,1\}$ is chosen uniformly and then a ciphertext $c \leftarrow \text{Enc}_k(m_b)$ is given to A.

4. A now has to guess, if he was given either m_0 or m_1. He continues to have access to $\text{Enc}_k(\cdot)$ and finally outputs $b' \in \{0,1\}$.

5. If $b' = b$ the output of the experiment is 1, else 0. For output 1 we say: A succeeds.

Now, if A is not able to succeed (or to guess right) with a probability significantly better than guessing, we call the encryption scheme Π *secure against chosen plaintext attacks*. This leads us to the following definition.

Definition 1. A private-key encryption scheme $\Pi = (Gen, Enc, Dec)$ has indistinguishable encryptions under a a chosen-plaintext attack, or is *CPA-secure*, if for all PPT adversaries A there is a negligible function $negl$ such that:

$$Pr\left[PrivK_{A,\Pi}^{cpa}(n) = 1\right] \leq \frac{1}{2} + negl(n)$$

where the probability is taken over the randomness used by A, as well as the randomness used in the experiment.

3 Pseudorandom Functions

As we now have accepted the importance of keeping our communication channel CPA-secure, naturally we now want to construct an encryption scheme secure against such chosen-plaintext attacks. To do that, we introduce so-called *pseudorandom functions*.

Pseudorandom functions are the generalization of *peusorandom generators*, we heard about in earlier talks, but instead of looking at "random-looking" strings, we now consider "random-looking" functions. Now, of course, it makes no sense, to call a *fixed* function $f : \{0,1\}^* \to \{0,1\}^*$ pseudorandom or even random, so we have to view *pseudorandomness* for the distribution on a set of functions. To do that, let me introduce the following definition.

Definition 2. A *keyed function* is a two input function $F : \{0,1\}^{\ell_{key}(n)} \times \{0,1\}^{\ell_{in}(n)} \to \{0,1\}^{\ell_{out}(n)}$ with $\ell_{key}(n), \ell_{in}(n), \ell_{out}(n) \in \mathbb{N}$, where we call the first input the key k. F is *efficient*, if there is a polynomial time algorithm that computes $F_k(x) := F(k, x)$ for arbitrary k, x.

For simplicity reasons, let us from now on assume that F is *length-preserving*, meaning that after picking a security parameter $n \in \mathbb{N}$ it holds that $\ell_{in}(n) = \ell_{out}(n) = \ell_{key}(n) = n$. So now, after fixing a key $k \in \{0,1\}^n$, we get a function F_k mapping n-bit strings to n-bit Strings which is the intuitive definition of the word *length-preserving*.

Consequently, we get get a natural distribution of functions from the above definition, which we call $\text{Func}_n := \{f : \{0,1\}^n \to \{0,1\}^n \mid f \text{ is a function}\}$. We can now laxly say that a function F is *pseudorandom*, if no polynomial time adverssary can distinguish, whether it is interacting with F_k with a uniform key k or with a uniform function $f \in \text{Func}_n$.

Since, when asking for security of an encryption scheme, we want to calculate probabilities, so let us take a quick glance at the cardinality of Func_n. We first notice that it holds that $|\{0,1\}^n| = 2^n$, since we have n-tuple with 2 possible entries in each line.

What is happening for any function f in the set Func_n is, that for every possible input $x \in \{0,1\}^n$ there are 2^n possible outputs, as we have calculated above. Of course, there are also 2^n possible inputs for the same reason. We can now visualize this with a combinatoric experiment: We draw 2^n balls (one for every possible output) from a pool of 2^n (one for every possible input) balls with putting them back. Luckily, the domain and the range are equivilant, so this is legit. Basic combinatorics gives us:

$$|\text{Func}_n| = |\{0,1\}^n|^{|\{0,1\}^n|} = (2^n)^{2^n} = 2^{n \cdot 2^n}$$

Through the above experiment, we also get that, when choosing $f \in \text{Func}_n$ uniformly, we could also pick 2^n entries from $\{0,1\}^n$ uniformly and then match an entry in $\{0,1\}^n$ to each. This also gives us for $x \neq y$ that $f(x)$ and $f(y)$ are independent.

Now, let us recall that F_k is a keyed function with $k \in \{0,1\}^n$. While F_k is chosen from 2^n distinct functions (since the key k is chosen from 2^n distinct values and F_k is clearly determined through k), f is chosen from $2^{n \cdot 2^n}$ functions, but they should still look the same to any efficient distinguisher. Formalizing this notion brings us to the following experiment.

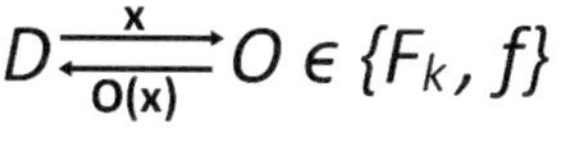

$$D \underset{O(x)}{\overset{x}{\rightleftarrows}} O \in \{F_k, f\}$$

Figure 3: The pseudorandom function distinguishing experiment

As we see in the above figure, a distinguisher D queries an Oracle O, which is either F_k or f for any value x. Here, D can freely interact with O and even adaptively chose x. D should be efficient, so D can only ask polynomially many questions.

Constructive to this experiment, below there is the formal definition for *pseudorandom functions*.

Definition 3. Let $F : \{0,1\}^* \times \{0,1\}^* \to \{0,1\}^*$ be an efficient, length-preserving, keyed function. F is a *pseudorandom function*, if for all PPT distinguishers D, there is a negligible function $negl$ such that:
$$|Pr\left[D^{F_k(\cdot)}(n) = 1\right] - Pr\left[D^{f(\cdot)}(n) = 1\right]| \leq negl(n)$$
where the first probability is taken over uniform choice of $k \in \{0,1\}^n$ and the randomness of D, and the second probability is taken over uniform choice of $f \in Func_n$ and the randomness of D.

Of course, in this scenario the key k is kept secret from the distinguisher D. Otherwise, any claims about the pseudorandomness of F_k no longer hold.
To understand, what a pseudorandom function might be, we look at an insecure example.

Example 3. Let F be $F_k(x) = x \oplus k$ a XOR-ing, keyed and length-preserving function with uniform k. It holds that for every input x the output $F_k(x)$ is uniformly distributed, yet F is not pseudorandom, since a distinguisher can query the oracle 2 distinct values x_1, x_2 to obtain $y_1 = O(x_1)$ and $y_2 = O(x_2)$. The distinguisher then outputs 1 (or in other words that O is a pseudorandom function), if $x_1 \oplus x_2 = y_1 \oplus y_2$, since it would then with the commutativity of $\oplus$ hold that

$$y_1 \oplus y_2 = F_k(x_1) \oplus F_k(x_2) = x_1 \oplus k \oplus x_2 \oplus k = x_1 \oplus x_2 \oplus k \oplus k = x_1 \oplus x_2 \oplus (0,0,...,0) = x_1 \oplus x_2$$

So, if $O = F_k$, the distinguisher has a success probability of 1. If O was a truly random function, since $f(x_1) \oplus f(x_2)$ has 2^n possible outcomes, it holds that

$$Pr\left[x_1 \oplus x_2 = f(x_1) \oplus f(x_2)\right] = 2^{-n}$$

We now see: The difference between those two probabilities is not negligible and therefore F cannot be a pseudorandom function

4 Pseudorandom Permutations

Except viewing arbitrary functions, we now require F to be a *permutation*, meaning that this $F_k : \{0,1\}^n \to \{0,1\}^n$ is a bijection. This naturally induces a distribution on the set of functions $\text{Perm}_n := \{f : \{0,1\}^n \to \{0,1\}^n \mid f\ bijective\}$.
With a similar combinatoric argument as above (only that we now draw balls from a pool without putting them back), we quickly calculate the cardinality for this set:

$$|\text{Perm}_n| = |\{0,1\}^n|! = (2^n)!$$

From this point of view, we can quickly transform our definitions for *pseudorandom permutations*. Of course, F is a *keyed permutation*, if $F : \{0,1\}^{\ell_{key}(n)} \times \{0,1\}^{\ell_{in}(n)} \to \{0,1\}^{\ell_{out}(n)}$ is a keyed function and additionally $F_k : \{0,1\}^{\ell_{in}(n)} \to \{0,1\}^{\ell_{out}(n)}$ is a bijection. Of course, this only makes sense, if we require $\ell_{in}(n) = \ell_{out}(n)$. Additionally, we assume F is *length-preserving*, meaning that $\ell_{key}(n) = \ell_{in}(n) = n$. We call $\ell_{in}(n) = n$ the *block length* of F.
We say that F is efficient, if there is a polynomial time algorithm to compute $F_k(x)$ as well as $F_k^{-1}(y)$ with given x, y and k.
With all this, it is obvious that for giving the definition of *pseudorandom permutations*, one must mostly copy the definition for *pseudorandom functions*, only substituting the uniform function with a uniform permutation. The the following remark shows us that this is nothing more, but a mere aesthetic choice, if the block length is sufficiently large.

Remark 1. *Let F be a pseudorandom permutation. Assuming $\ell_{in}(n) \geq n$, it holds that F is also a pseudorandom function.*

Proof: Let us without loss of generality assume that $\ell_{in}(n) = n$. We'll later see, why this is possible. Let F be a pseudorandom function and let $f \in \text{Perm}_n$ be uniform randomly chosen. Let D be an efficient distinguisher. Per definition it holds that

$$|Pr\left[D^{F_k(\cdot)}(n) = 1\right] - Pr\left[D^{f(\cdot)}(n) = 1\right]| \leq negl(n)$$

for some negligible function $negl$. Let $q \in \mathbb{R}[X]$ be a polynomial for the efficient distinguisher D that works as an upper bound for the number of queries, D executes to its oracle. Let's now randomly chose a uniform $g \in \text{Func}_n$. Now, the only new options, D has to determine, whether he is looking at g or F_k is to find two values x, y so that $g(x) = g(y)$, since in that case, it could not be a permutation. How many "new" cases are that?
There are 2^n possible values in the range of g. There are several options that need to be viewed:

1. Every value in the image of g is the same. That makes up 2^n possibilities for every possible value in the range.

2. Every value in the image of g is the same except for one. That makes up $2^n \cdot \binom{n}{1} \cdot (2^n - 1)$ possibilities, since we have 2^n entries in the domain to chose the same value, $2^n - 1$ values in the domain to chose the distinct value and we can order the values the way we want, hence $\binom{n}{1}$.

3. For arbitrary $2 \leq k \leq n$ congruent values we now get with this idea $2^n \cdot \binom{n}{k} \cdot \frac{(2^n-1)!}{(2^n-n+k-1)!}$.

All of this combined gives us, where q is some polynomial, since D is efficient:

$$\begin{aligned}
Pr[A] := Pr[g(x) = g(y) \; for \; x \neq y] &= q(n) \cdot \frac{2^n}{2^{n \cdot 2^n}} \cdot \sum_{k=2}^{n} \binom{n}{k} \cdot \frac{(2^n - 1)!}{(2^n - n + k - 1)!} \\
&\leq q(n) \cdot \frac{2^n}{2^{n \cdot 2^n}} \cdot \underbrace{\sum_{k=1}^{n} \binom{n}{k}}_{=2^n - 1} \cdot \frac{(2^n - 1)!}{(2^n - n - 1)!} \\
&\leq \frac{q(n)}{2^{n \cdot (2^n - 2)}} \cdot (2^n)^n \\
&= \frac{q(n)}{2^{n \cdot (2^n - 2) - n^2}} =: K(n)
\end{aligned}$$

Now, $K(n)$ is negligible, because 2^n grows faster than n^2. We therefore define $negl'(n) := K(n)$. This, and the fact that the sum of two negligible functions itself is negligible, gives us:

$$\begin{aligned}
|Pr\left[D^{F_k(\cdot)}(n) = 1\right] - Pr\left[D^{g(\cdot)}(n) = 1\right]| &\leq |Pr\left[D^{F_k(\cdot)}(n) = 1\right] - Pr\left[D^{f(\cdot)}(n) = 1\right]| + Pr[A] \\
&\leq negl(n) + negl'(n) =: negl''(n) \\
\Rightarrow F \; is \quad a \quad &pseudorandom \; function
\end{aligned}$$

We now observe that we could (for the case $\ell_{in}(n) > n$) just gauge the Term $K(n)$ upwards with a term $K'(n)$, where $\ell_{in}(n)$ is substituted by n and the rest of the inequations would still hold, meaning that $K'(n)$ is also a negligible function. This is possible, because $K(n)$ is negligible and therefore it holds that the denominator is far greater than the numerator for big n, e.g. $2^n \gg q(n)$. Since this also holds for big $\ell_{in}(n)$ and $\ell_{in}(n) \geq n$ is assumed, it holds that

$$\frac{q(n)}{2^{n \cdot (2^n - 2) - n^2}} \geq \frac{q(\ell_{in}(n))}{2^{\ell_{in}(n) \cdot (2^{\ell_{in}(n)} - 2) - \ell_{in}(n)^2}}$$

Which justifies our w.l.o.g. assumption from the start. □

Now, if F is a keyed permutation, cryptographic schemes based on F might require the honest parties to compute F_k^{-1} in addition to F_k. So that this does not influence the security of the scheme, we demand F_k to be indistinguishable from a uniform permutation, even if the distinguisher is additionally given oracle access to the inverse of the permutation. We then call F a strong pseudorandom permutation, as the following definition states.

Definition 4. Let $F : \{0,1\}^* \times \{0,1\}^* \to \{0,1\}^*$ be an efficient, length-preserving, keyed function. F is a *strong pseudorandom permutation*, if for all PPT distinguishers D, there is a negligible function *negl* such that:

$$|Pr\left[D^{F_k(\cdot), F_k^{-1}(\cdot)}(n) = 1\right] - Pr\left[D^{f(\cdot), f^{-1}(\cdot)}(n) = 1\right]| \leq negl(n)$$

where the first probability is taken over uniform choice of $k \in \{0,1\}^n$ and the randomness of D, and the second probability is taken over uniform choice of $f \in Perm_n$ and the randomness of D.

A *strong pseudorandom permutation* is also called a *block cipher* (compare *stream cipher*). Of course it holds that any *strong pseudorandom permutation* is also a *pseudorandom permutation*. Also, expectedly, we can easily construct *pseudorandom generators* (which we have talked about the last week) from *pseudorandom functions*, for example (with arbitrary $l \in \mathbb{N}$):

$$G(s) := F_s(1) || F_s(2) || \ldots || F_s(l)$$

If we replaced F_s with a uniform function, the output of G would be uniform as well and therefore, when using F instead, the output would be pseudorandom.
Also, *block ciphers* can be used to construct *stream ciphers*. Now, whereas stream ciphers typically have a better performance, they a far less well-understood and therefore, there is a low confidence in their security. Thus, whenever possible, it is usually recommended to use block ciphers, for example by converting them to stream ciphers first.

Now we could hope that we solved the problem from last chapter that it is really hard to find *pseudorandom generators*. Unfortunately, this problem resumes for pseudorandom functions: We do not know any pseudorandom functions, yet we assume their existance to prove security of encryption schemes built with them.
With this in mind, in the following we are trying to construct a *CPA-secure* encryption scheme, assuming that pseudorandom functions exist.

5 CPA-Secure Encryption

As observed earlier, an encryption scheme could not possibly be *CPA-secure*, if it was deterministic, meaning the same input x would always yield the same encryption $Enc_k(x)$, for example by simply defining $Enc_k(x) := F_k(x)$.

To compensate this, naturally we randomize our encryption. We do that by first chosing a uniform $r \in \{0,1\}^n$ and then applying our pseudorandom function F to this value, giving us $F_k(r)$ which we combine with the plaintext message $m \in \{0,1\}^n$ by XOR-ing them. Consequently, as the other communication party requires r to decode the message, the output of the encryption is the ciphertext $c = (r, F_k(r) \oplus m)$. The below figure illustrates, how this works. In the following, I

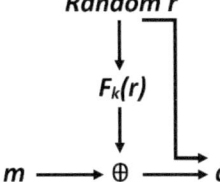

Figure 4: A *CPA-secure* encryption scheme

try to formalize, what was said above.

Constructing a CPA-secure encryption scheme:

Let F be a pseudorandom function. We define a private-key encryptio scheme Π for arbitrary messages of length n as follows:

1. Gen: on input n generates a uniform $k \in \{0,1\}^n$ and outputs it.

2. Enc: given a key $k \in \{0,1\}^n$ and a message $m \in \{0,1\}^n$, we choose a uniform $r \in \{0,1\}^n$ and output the ciphertext
$$c := (r, F_k(r) \oplus m)$$

3. Dec: given $k \in \{0,1\}^n$ and a ciphertext $c = (r, s)$, output the plaintext message
$$m := F_k(r) \oplus s$$

First, let us confirm that this encryption scheme is correct, in other words that it holds that $Dec_k(Enc_k(m)) = m$. For that matter we simply calculate, what is being given, only using the associativity of $\oplus$:

$$Dec_k(Enc_k(m)) = Dec_k((r, F_k(r) \oplus m)) = F_k(r) \oplus (F_k(r) \oplus m) = (0, 0, \ldots, 0) \oplus m = m$$

To gain further understanding of this, we are going to look at the following (insecure) example with the *non-pseudorandom function* having the values $F_k(x) = k \oplus x$. Let us fix the following: $n = 8, m = 10100011, k = 10001101$. We then pick a uniform $r \in \{0,1\}^8$, so let $r = 10101010$. How should this be encrypted by our scheme?

$$Enc_k(m) = (r, F_k(r) \oplus m) = (10101010, 10001101 \oplus 10101010 \oplus 10100011) = (10101010, 10000110)$$

This ciphertext is now sent to another honest party, which then tries to decrypt it, giving us:

$$Dec_k(10101010, 10000110) = F_k(10101010) \oplus 10000110 = 10100011 = m$$

Now that we have seen the correctness of this scheme and have understood, how it works by looking at an example, we want to prove that it in fact is CPA-secure, using our proof strategy from the appendix. We therefore give the following theorem.

Theorem 1. *If F is a pseudorandom function, then the above Construction is a CPA-secure private-key encryption scheme for messages of length n.*

Proof: Let $\tilde{\Pi} = (\tilde{Gen}, \tilde{Enc}, \tilde{Dec})$ be an encryption scheme that is exactly the same as the encryption scheme from the above construction, but uses a uniform function f instead of a pseudorandom function. Let us now fix an arbitrary probabilistic polynomial time adversary A with $q \in \mathbb{R}[X]$, where $q(n)$ is the upper bound for queries, A makes to its oracle, given the input n.

We now use A to construct a distinguisher D with the aim to determine, whether its oracle function is pseudorandom or truly random.

Distinguisher D: D is given input n and access to an Oracle O.

1. Run $A(n)$. Whenever A queries O for $m \in \{0,1\}^n$, answer:
   - Choose uniform $r \in \{0,1\}^n$
   - Query $O(r) =: y$
   - Return $(r, y \oplus m)$ to A

2. When A outputs $m_0, m_1 \in \{0,1\}^n$, choose uniform bit b and:
   - Choose uniform $r \in \{0,1\}^n$
   - Query $O(r) =: y$
   - Return $(r, y \oplus m_b)$ to A

3. Repeat 1. until A outputs a bit b'. Output 1 if $b' = b$, 0 otherwise.

Obviously, D runs in polynomial time, since A does. We now notice for the two cases that O is either a pseudorandom function or a random function:

<u>Case 1 $[O = F_k]$</u>: A key k is chosen uniformly at random and then for every encryption r is chosen uniformly for computing $y := F_k(r)$ and setting $c = (r, y \oplus m)$, exactly like in the above construction. With this in mind, in the view of A run as a subroutine of D, it is happening exactly the same as in the experiment $\text{PrivK}_{A,\Pi}^{cpa}$. It therefore holds that

$$Pr_{k \leftarrow \{0,1\}^n}\left[D^{F_k(\cdot)}(n) = 1\right] = Pr\left[\text{PrivK}_{A,\Pi}^{cpa}(n) = 1\right] =: Pr\left[cpa\right]$$

<u>Case 2 $[O = f \in Func_n]$</u>: An analogous argument for a uniform function $f \in Func_n$ and the experiment from an earlier section $\text{PrivK}_{A,\tilde{\Pi}}^{cpa}$ gives us

$$Pr_{f \leftarrow Func_n}\left[D^{f(\cdot)}(n) = 1\right] = Pr\left[\text{PrivK}_{A,\tilde{\Pi}}^{cpa}(n) = 1\right] =: Pr\left[\tilde{cpa}\right]$$

Since F is pseudorandom, we also know that for some negligible function $negl$ it holds that

$$\left|Pr\left[D^{F_k(\cdot)}(n) = 1\right] - Pr\left[D^{f(\cdot)}(n) = 1\right]\right| \leq negl(n)$$

All of the above combined gives us

$$Pr\,[cpa] - Pr\,[c\tilde{p}a] \leq |Pr\,[cpa] - Pr\,[c\tilde{p}a]| = |Pr\left[D^{F_k(\cdot)}(n) = 1\right] - Pr\left[D^{f(\cdot)}(n) = 1\right]| \leq negl(n)$$

And therefore

$$Pr\,[cpa] \leq Pr\,[c\tilde{p}a] + negl(n) \qquad (1)$$

Now, in the next step let r^* be the distinct value used to create the ciphertext $(r^*, f(r^*) \oplus m_b)$ (compare to the construction of the Distinguisher D). There are two possibilities, what can happen.

REP: the value r^* is used at least once, when answering any of A's encryption oracle queries. In this case, A can determine whether m_0 or m_1 have been encryted with a probability of 1, because if the oracle ever answers with a ciphertext (r^*, s), A learns that $f(r^*) = s \oplus m$.
However, A is efficient with at most $q(n)$ queries and r^* is chosen from 2^n distinct values, so the probability of this event is at most $q(n)/2^n$, which is negligible.

$\neg$ REP: the value r^* is never used, when answering any of A's encryption oracle queries. Since $f(r^*)$ is uniformly distributed, A learns nothing about the encryption scheme and therefore succeeds with a probability $1/2$.

Combining this with the definition of conditional probabilities $Pr\,[A \cap B] = \overbrace{Pr\,[B]}^{\leq 1} \cdot Pr\,[A \mid B]$ and the fact $Pr[A] = Pr[A \land \neg B] + Pr[A \land B]$ gives us:

$$\begin{aligned} Pr\,[c\tilde{p}a] &= Pr\,[c\tilde{p}a \land REP] + Pr\,[c\tilde{p}a \land \neg REP] \\ &\leq Pr\,[REP] + Pr\,[c\tilde{p}a \mid \neg REP] \\ &\leq \frac{q(n)}{2^n} + \frac{1}{2} =: negl'(n) + \frac{1}{2} \end{aligned}$$

This and the inequation (1) finally gives us, since the sum of two negligible functions again is a negligible function:

$$Pr\,[cpa] \leq Pr\,[c\tilde{p}a] + negl(n) \leq negl'(n) + \frac{1}{2} + negl(n) =: negl''(n) + \frac{1}{2}$$

Thus, when looking at the definition of *CPA-security*, with this we see that Π is a CPA-secure encryption scheme. $\square$

6 Literature

Katz, Jonathan and Lindell, Yehuda (2007): Introduction to Modern Cryptography, 2nd Edition. Chapman & Hall/CRC, Boca Raton, FL

Boaz, Barak (2007): Lecture 5 - CPA security, Pseudorandom functions.

7 Appendix

Proofs of security based on pseudorandom functions.
In this section I want to give a template that is used in most proofs of security for encryption schemes containing *pseudorandom functions*. In the first step, we consider a hypothetical version of the encryption scheme, where the pseudorandom function is replaced with a random function. It is then argued that this modification does not significantly affect the adversary's success probability. After that, we are left with analyzing a scheme that uses completely random function. From this point on, the proof relies on probabilistic calculus rather than on computational assumptions, finishing the proof.

YOUR KNOWLEDGE HAS VALUE

- We will publish your bachelor's and master's thesis, essays and papers

- Your own eBook and book - sold worldwide in all relevant shops

- Earn money with each sale

Upload your text at www.GRIN.com
and publish for free